Animal Behavior Investigations

ANIMAL COMMUNICATION

By Michelle Garcia Andersen

Rourke Educational Media

A Division of Carson Dellosa Education

BEFORE AND DURING READING ACTIVITIES

Before Reading: *Building Background Knowledge and Vocabulary*

Building background knowledge can help children process new information and build upon what they already know. Before reading a book, it is important to tap into what children already know about the topic. This will help them develop their vocabulary and increase their reading comprehension.

Questions and Activities to Build Background Knowledge:

1. Look at the front cover of the book and read the title. What do you think this book will be about?
2. What do you already know about this topic?
3. Take a book walk and skim the pages. Look at the table of contents, photographs, captions, and bold words. Did these text features give you any information or predictions about what you will read in this book?

Vocabulary: *Vocabulary Is Key to Reading Comprehension*

Use the following directions to prompt a conversation about each word.

- Read the vocabulary words.
- What comes to mind when you see each word?
- What do you think each word means?

Vocabulary Words:
- *antennae*
- *bond*
- *communication*
- *pheromones*
- *poisonous*
- *precise*
- *scent*
- *signals*

During Reading: *Reading for Meaning and Understanding*

To achieve deep comprehension of a book, children are encouraged to use close reading strategies. During reading, it is important to have children stop and make connections. These connections result in deeper analysis and understanding of a book.

Close Reading a Text

During reading, have children stop and talk about the following:

- Any confusing parts
- Any unknown words
- Text to text, text to self, text to world connections
- The main idea in each chapter or heading

Encourage children to use context clues to determine the meaning of any unknown words. These strategies will help children learn to analyze the text more thoroughly as they read.

When you are finished reading this book, turn to the next-to-last page for **Text-Dependent Questions** and an **Extension Activity**.

Table of Contents

Case Study: What's That Smell?

Do you wish animals could talk? What do you think they would say? Animals don't speak like we do, but they do use **communication**. Even ants communicate, but how?

Florida Carpenter Ants

Scientists took ants from different nests. They made the ants fall asleep in a box. Next, they pumped chemicals that smelled like other ants into the box. The experts watched the sleeping ants closely. What do you think they were looking for?

When the **scent** of ants from their own nests was used, the ants' **antennae** moved one way. When the scent came from ants who were "strangers," the ants' antennae moved differently. Why do you think this happened? As you learn more about animal communication, see if you can solve this mystery.

In the Air

We use more than our noses to smell. When we sniff, we pick up bits of scent floating in the air. We send that information to our brains. Our brains tell us what we're smelling.

Are You Talking to Me?

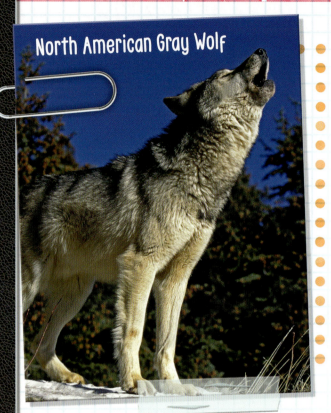

North American Gray Wolf

Communication is how we share information. Animals communicate with **signals**. Signals can be seen, smelled, heard, or felt.

Animals usually, but not always, send signals to the same kind of animals. For example, wolves communicate with each other by howling. Koko was a western lowland gorilla who used sign language, a kind of signal that can be seen, to communicate with people. She knew many words. She communicated as well as a three-year-old person using these signals.

Animals also communicate with other kinds of living things. Have you ever smelled a skunk? They are able to spray an unpleasant liquid when they are scared. It tells other animals to leave them alone.

Western Lowland Gorilla

Koko's Kitten

Koko once asked for a kitten for Christmas. Instead, she was given a toy kitten. She signed the word "sad" and didn't play with the toy. On her birthday, she was given a real kitten. Koko treated him like her baby.

Sound is important for communication, too. Rattlesnakes shake their tails to warn others to back off. Tiger moths make clicking sounds. These sounds warn hungry bats that the moths don't taste good.

Western Diamondback Rattlesnake

Little Brown Bat

Tiger Moth

Giant Root-Rat

Other animals send messages that can be felt. This includes thumping and tapping, like when giant root-rats "talk" by banging their heads on tunnels underground.

More Than Meets the Eye

Animals communicate for many reasons: to find mates, for protection, to share information in a group, and to care for their young.

Male peacocks like to show off by fanning and waving their tail feathers. This makes them look bigger. Showing off is how they find mates.

Peacock and Peahen

Green-and-Black Poison Dart Frog

Poisonous dart frogs are colorful and bright. Their colors warn other animals not to eat them. They can be deadly, even to bigger animals.

Red-Backed Poison Dart Frog

Vervet monkeys use sounds when they have something to say. Their calls are very <mark>precise</mark>. If they see a hunting snake, they tell the other monkeys to hide in trees. If they see a dangerous eagle, their call warns the others to crouch down.

Crowned Eagle

Vervet Monkey

African Rock Python

Whales also make sounds to communicate. They whistle, click, and make pulse noises. Whales use sound to find food and other whales.

Whale Songs

Do you like to sing? So do humpback whales! The male humpback whale sings songs to attract females.

Fireflies

Some animals communicate with chemicals. The firefly makes a chemical so its body can glow. When it's young, it glows as a warning that it doesn't taste good. When it is older, it will glow to find a mate. Fireflies flash their lights and glow to find others.

Canada Lynx

Some animals make chemicals called **pheromones**. Pheromones leave a scent. This is a signal to animals of the same kind. A female Canada lynx leaves behind a scent when she urinates and claws at trees. Other cats can smell her and stay away. This helps protect her kittens.

Many animals communicate with touch. Elephants use touch to **bond**. They use their trunks to show their feelings. They also communicate with sounds by trumpeting and stomping. Some of their sounds cannot be heard by humans.

Honey bees communicate by touch. They do a dance called the waggle dance. It lets the other bees know where to find food. When a bee dances inside her dark hive, the other bees can't see the dance. The bees can feel the dance through their feet.

Honey Bees

Elephants

Think about the ants in the study at the beginning of the book. Which form of communication do you think they use? Why do you think this happens?

Ant Queen

If you said they use their sense of smell, you are right! Ants use their antennae to smell. The study showed that ants can smell body odor. They know which ants belong to their nest by their scent. Ants have jobs inside their nest. They can tell which job each ant does by the ant's scent. They know who the queen ant is by her smell. They know who the worker ants are, too. This helps keep order in their nest.

To learn more about the case study and scientists in this book, search online for "Dr. Kavita Sharma" and "*Camponotus floridanus.*"

Get Creative! Animal Extension Activity

Animals communicate with each other in many ways, including using sounds, vibrations, and lights. Some animals use flash patterns, and others dance to get their message through. Design an experiment to test a new, animal-inspired way that you can communicate with your friends.

1. Brainstorm three different and new ways that you could send messages back and forth between your friends.

2. Make a list of the pros and cons of each method.

3. Decide how your new communication method will work for letters, sounds, words, or even sentences.

4. Test your communication by having a conversation with a friend!

Design your study. Make sure it is safe for living things. Write a plan for your study and share it with a friend.

Glossary

antennae (an-TEN-uh): feelers on the head of an insect

bond (bond): to form a close relationship with

communication (kuh-myoo-ni-KAY-shuhn): the act of sharing information, ideas, and feelings with others

pheromones (FEH-ruh-mownz): a scent given off by some animals

poisonous (POI-zuh-nuhs): having a poison that can harm or kill

precise (pri-SISE): very exact or with great attention to detail

scent (sent): a distinctive smell or odor

signals (SIG-nuhls): signs that send messages or warnings

Index

Text-Dependent Questions

1. Describe four ways that animals communicate.
2. Why are poisonous dart frogs so brightly colored?
3. What are pheromones, and what do they do?
4. What is the name of the dance the honey bee does?
5. Think about the many animals you read about in this book and how they communicate. Which did you find most interesting and why?

Further Reading

Rake, Jody S. *The Naked Mole-Rat*, Capstone, 2008.

Simon, Seymour. *Elephants*, HarperCollins, 2018.

Socha, Piotr. *Bees*, Abrams, 2017.

About the Author

Michelle Garcia Andersen lives in southern Oregon, surrounded by pets and wildlife. One of Michelle's favorite pastimes is sitting on her front porch observing nature and those who inhabit it. She feels most grateful when the elk come into view and she can take time out of her day to listen and watch them communicate with their herd.

www.rourkeeducationalmedia.com

PHOTO CREDIT: Cover ©Waffa, ©TippaPatt, ©Olha Kozachenko, ©sruilk; back cover ©Mikhail Semenov; p3 ©32 Pixelsp4 ©mprikaznov; p5 ©Dgrilla; p6 ©Russell Marshall; p7 ©Russell Marshall; p8 ©slowmotiongli; p9 ©Dirk M. De Boer, ©32 Pixels; p10 ©DMartin09; p11 ©Michael Durham, ©Kirsanov Valery Vladimirovich, ©Giedriius; p12 ©Lucia Gajdosikova; p13 ©Dirk Ercken; p14 ©nwdph, ©Earnie Janes, ©Cesar J. Pollo, ©32 Pixels; p14 ©32 Pixels; p15 ©Fer Gregory, ©32 Pixels; p16 ©Reimar, ©32 Pixels; p17 ©Diyana Dimitrova, ©Nick Greaves; p18 ©thrall Shula; p19 ©Pavel Krasensky; p20 ©Russell Marshall; p21 ©BonD80, ©Boltenkoff, ©Evgeny Karandaev, ©Hurst Photo, ©Frame Art; p24 ©Michelle Garcia Andersen

Edited by: Tracie Santos
Cover and interior layout by: Tammy Ortner

Library of Congress PCN Data

Animal Communication / Michelle Garcia Andersen
(Animal Behavior Investigations)
ISBN 978-1-73164-937-9 (hard cover)(alk. paper)
ISBN 978-1-73164-885-3 (soft cover)
ISBN 978-1-73164-989-8 (e-Book)
ISBN 978-1-73165-041-2 (ePub)
Library of Congress Control Number: 2021935267

Rourke Educational Media
Printed in the United States of America
01-1872111937